Instruction Manual to Accompany *Elements of Molecular Neurobiology* *(Second Edition)*

Instruction Manual to Accompany

Elements of Molecular Neurobiology

(Second Edition)

C. U. M. SMITH

Department of Vision Sciences,
Aston University, Birmingham, UK

and

H. RICHARD LEUCHTAG

Department of Biology,
Texas Southern University, Houston, Texas, USA

JOHN WILEY & SONS

Chichester · New York · Weinheim · Brisbane · Toronto · Singapore

National 01243 779777
International (+44) 1243 779777
e-mail (for orders and customer service enquiries): cs-books@wiley.co.uk
Visit our Home Page on http://www.wiley.co.uk
or http://www.wiley.com

Other Wiley Editorial Offices

John Wiley & Sons, Inc., 605 Third Avenue,
New York, NY 10158-0012, USA

WILEY-VCH Verlag GmbH, Pappelallee 3,
D-69469 Weinheim, Germany

Jacaranda Wiley Ltd, 33 Park Road, Milton,
Queensland 4064, Australia

John Wiley & Sons (Asia) Pte Ltd, Clementi Loop #02-01,
Jin Xing Distripark, Singapore 129809

John Wiley & Sons (Canada) Ltd, 22 Worcester Road,
Rexdale, Ontario M9W 1L1, Canada

British Library Cataloguing-in-Publication Data

A catalogue record for this book is available from the British Library

ISBN 0-471-97543-5

Typeset in 10/11½pt Times from the author's disks by Dobbie Typesetting Ltd, Tavistock, Devon
Printed and bound in Great Britain by RPM Reprographics, Chichester
This book is printed on acid-free paper responsibly manufactured from sustainable forestation, for which at least two trees are planted for each one used for paper production.

CONTENTS

PREFACE

Molecular neurobiology is a deep and complex subject. It requires knowledge of almost every branch of science. In addition to its intrinsic interest, it forms the basis for many other parts of neuroscience and neurology. With advances coming at breakneck pace, it is easy for new students to lose their way in the mass of detail pouring from the research laboratories. This book originated in the search for a way to teach a course as challenging as molecular neurobiology to groups of students of varying backgrounds and attainments. Part of the answer, we felt, had to be frequent feedback. Eventually a list of questions, designed to help the students focus their study efforts, was generated. It was linked to the textbook, *Elements of Molecular Neurobiology* by C. U. M. Smith, providing a set of questions for each chapter of the text.

This *Instruction Manual* is the product of an international collaboration by e-mail between its two authors. It is intended both for course instructors and students. It is presented to instructors as a tool to assist them in directing students along fruitful avenues of thought about the molecules of the brain, itself the organ of thought. The brief summaries provide nutshell orientations to the chapters. The question sequence generally follows the order in which the material is presented. The questions are designed to attract the student's attention to the salient points of the text, with each section (and each box) being tested with at least one question.

Of course, well-prepared and disciplined students may have little need for a manual of this type. Nevertheless, there are few who cannot benefit from the opportunity which this manual provides to focus their intellectual powers on a complex and exciting field of study. The essay-type questions provide a varied source of feedback to the textual material, although deliberately formulated in language different from the original presentation. Some questions are factual; others are 'thought questions' that test a student's ability to apply principles to new examples or to combine information from different places. A review of electrical concepts, *Current, Voltage and All That*, may be helpful in mastering the material on the all-important membrane potentials and ion flows and the instruments to measure them.

By carefully working through this manual in association with the text, students should achieve a thorough grasp of the elements of molecular neurobiology. *Elements of Molecular Neurobiology* aims to provide students with a grounding and a framework within which to build their understanding. Having mastered the basics and understood the major themes and problems, students are in a position to use their own initiative. To this end, the text provides a full bibliography and advice on where to look for yet more recent knowledge.

The authors gratefully acknowledge the helpful comments of the reviewers and the help of the Wiley editors, especially Janie Curtis, Publisher. We would be delighted to receive your comments on your experiences with this book.

If by any chance you have this instruction manual but not the main text, you should be able to purchase a copy at any good academic booksellers. If you have trouble obtaining a copy of *Elements of Molecular Neurobiology (Second Edition)*, please call John Wiley & Sons, Ltd on (+44) 1243 779777 or John Wiley & Sons, Inc. on (+1) 212 850 6000.

1

INTRODUCTORY ORIENTATION

Some animals have no nervous system; others lack a central nervous system. The roundworm *C. elegans* has 302 well-mapped neurons, while the enormous number of neurons in the human central nervous system far exceeds the number of genes or the capacity of the genome to determine their connections. The embryonic development of an individual vertebrate brain appears to retell the story of its evolution. The complexity of the brain rests on the manifold forms of only two types of cells: neurons and glia. The brain is protected by its three coverings, the outermost dura mater, the arachnoid and the pia mater. Exquisitely varied patterns of synapses connect neuron to neuron. The complexity of connections in the cerebral cortex barely begins to be described by its layers and modules.

1.1 How does the design of the CNS of chordates differ from that of invertebrates?

1.2 Describe the embryological development of the vertebrate brain up to the five-vesicle stage.

1.3 How does the upright stance of *Homo sapiens* affect brain development?

1.4 Where is the limbic system located? What is its functional significance?

1.5 How are monopolar, bipolar and multipolar neurons distinguished? How are they similar?

1.6 Sketch and describe a typical multipolar neuron.

1.7 What is axoplasmic flow? What structure does it involve? In what direction(s) does it run?

1.8 Name and describe the functions of the three types of glia.

1.9 Sketch and describe a classical synapse. What other types of synapses are there?

1.10 Approximately what proportion of neurons does the Golgi–Cox silver staining technique affect?

1.11 Distinguish between grey and white matter. Describe the composition or consistency of these tissues that earn them their respective names.

1.12 How does the arrangement of the grey matter of the brain change in the course of human embryonic development? To what new structures does this change lead?

1.13 Describe the stratification of the neocortex. Where do most of the axons leaving the cortex go?

2

THE CONFORMATION OF INFORMATIONAL MACROMOLECULES

The information that resides in the brain ultimately depends on information that resides in its molecules. Proteins, built up linearly from a selection of twenty amino acids, form the dynamic structures that bring about the cellular events of brain activity. Nucleic acids carry the continuity of life in the genome and build protein enzymes and structures from the genetic plan.

2.1 What are the informational macromolecules? Why is that term appropriate?

2.2 What properties of amino acids have particular significance in determining protein structure? What is meant by the terms 'hydrophobic' and 'hydrophilic'?

2.3 What is the primary structure of a protein? What bonds hold it together? What are the terminal groups?

2.4 A tripeptide has three amino acids. How many possible primary structures can it have? Explain.

2.5 What is the name of the process by which oligosaccharide side chains are attached to a protein? By what types of linkages are they attached?

2.6 What type of bonding accounts for the 'stickiness' of complementary surfaces? Describe it. How strong is it?

2.7 With reference to the secondary structure of proteins, explain what is meant by the terms 'parallel' and 'antiparallel'. Which has the more precise alignment? Give an example.

2.8 Describe some configurations of acetylcholinesterase.

2.9 Which amino acid can not be accommodated in an α-helix? How are the hydrogen bonds formed in an α-helix?

2.10 What feature does an α-helix in which every seventh residue is non-polar show? How do two such helices interact? What structures are formed in this way?

2.11 What is hydropathy analysis? What does it tell us about the structure of a transmembrane protein?

2.12 What forces determine the tertiary structure of a protein? About how strong are they in comparison to covalent bonds? What happens to the protein when these bonds are broken by heat?

2.13 What is the name for the structure of a multimeric protein? What holds it together? How strong are these forces?

2.14 Describe some of the 'structural motifs' or 'domains' found in globular proteins.

2.15 What roles do conformational changes play in the foundations of neurobiology?

2.16 What are molecular chaperons and why are they important?

2.17 Draw a diagram showing the molecular structure of a polynucleotide. Explain what is meant by the 5′ to 3′ direction.

2.18 Do the two polynucleotide strands in DNA align against each other in a parallel or antiparallel sense? Compare your answer to that for Question 2.7.

2.19 Name the three major types of RNA found in cells and briefly indicate their roles. Why is the three-dimensional structure of tRNA so intricate?

2.20 The nuclei of neurons can usually be distinguished from those of glial cells by the presence of one or more well-developed nucleoli. What part do nucleoli play in the life of a cell and why should they be more prominent in neurons than in glia?

3

INFORMATION PROCESSING IN CELLS

The interaction of nucleic acids with proteins is based on the genetic code, which assigns amino acids to triplets of nucleotides. Not all genes are expressed in a given cell, nor are all stretches of DNA meaningful. The expression of genetic information is controlled at the genomic, transcriptional, post-transcriptional, translational and post-translational levels.

3.1 What is the 'central dogma' of molecular biology? Cite an exception to it. How does it reinforce Darwinian concepts?

3.2 Name two amino acids for which the genetic code has no degeneracy.

3.3 Which strand of DNA is transcribed to mRNA? What is a function of the other strand?

3.4 Name two specific transcription factors and explain why they are crucial in the life of a neuron.

3.5 What is a palindromic sequence? Where are such sequences found?

3.6 In what way is transcription more complex in a neuron than in a prokaryote?

3.7 Are exons, introns or both transcribed? Which remain in the finished mRNA?

3.8 How can errors in post-translational processing lead to human hereditary diseases? Give examples.

3.9 What is meant by mRNA editing? Give a neurobiological example where it is important.

3.10 Describe the function of aminoacyl tRNA synthetase enzymes. Explain why they have been said to constitute the cell's second genetic code.

3.11 Draw a diagram to show how mRNA, tRNA and the ribosome interact in the synthesis of a polypeptide chain.

3.12 What aspect of the structure of a polypeptide is specified by the genetic code? What aspect is affected by environmental interactions? How may molecular chaperons be involved?

3.13 Explain how antisense oligonucleotides can be used to block the translation of mRNA and thus prevent the expression of a deleterious gene.

3.14 What determines whether a protein that has been synthesised will end up in the axoplasm, the membrane or the perikayon?

3.15 What is the relation between proto-oncogenes and oncogenes? Name two proto-oncogenes that are significant in neurobiology.

3.16 Describe the relationships between proto-oncogenes, immediate early genes (IEGs), transcription factors, palindromic sequences and neuropeptides.

3.17 With the aid of examples, describe how differential post-transcriptional processing generates different peptides in different tissues.

3.18 Compare and contrast post-translational processing with post-transcriptional processing. What is the biological significance of both these processes?

4

MOLECULAR EVOLUTION

The genetic plan is not immutable. Organisms compete with each other for resources. Innumerable different niches are filled and many different life-styles evolve. The evolutionary process generates a thicket of branching paths. Mutations lead to new characteristics, which are seized on by environmental selection. Within the genome genes are sometimes duplicated and transposons carry them from one location to another. The slow inexorable drip of point mutation causes proteins to change over geological time, and selection refines their structure and functioning. Knowing these mechanisms, the molecular geneticist is increasingly able to decipher evolutionary relationships.

4.1 What are the major components of Darwin's theory of evolution? How has molecular biology supported it?

4.2 Why is the brain particularly exposed to Darwinian selection?

4.3 What did Alfred Romer mean when he proposed that evolutionary advances were at first 'conservative' or 'reactionary'?

4.4 Explain the difference between 'homology' and 'analogy'.

4.5 How has the evolutionary approach helped advance molecular biology and neurobiology?

4.6 What are the three ways in which spontaneous point mutations may occur? Give examples of each type.

4.7 What types of mutations will cause only minor changes in the expressed protein? Give examples.

4.8 Distinguish between 'conservative' and 'radical' amino acid substitutions. Give examples.

4.9 What type of mutation preserves the number of rings in the deoxynucleotide base? What type changes the number? Which of these types of mutation is the more common? Explain.

4.10 What enzymes operate to proof-read DNA in prokaryocytes and in eukaryocytes? How is this carried out?

4.11 Referring to Figure 4.3, describe the process of 'nick translation'.

4.12 Which nucleotide base has the greater molecular weight, thymine or uracil? Cite an evolutionary advantage for the presence of thymine rather than uracil in DNA.

4.13 With the aid of a diagram, show how gene duplications can occur. Is this an important phenomenon in the evolution of proteins? Why?

4.14 What is 'exon shuffling'? How is it involved in the evolution of proteins?

4.15 What is the importance of knowing the rate at which proteins evolve? Do all proteins evolve at the same rate?

4.16 With reference to the globin molecule, discuss some of the major features of protein evolution. Is there any evidence to suggest that exon shuffling has occurred?

4.17 In what ways are the evolution of the globins and the nAChRs similar? (Refer also to Section 9.1.)

4.18 Provide definitions of 'families' and 'superfamilies' of proteins.

4.19 What is a polyprotein? Give three examples. Can the same precursor peptide yield different sets of end products in different tissues? Explain.

4.20 What is the evolutionary significance of polyproteins?

5

MANIPULATING BIOMOLECULES

Molecular biology has exerted and continues to exert a profound influence on the course of research in neurobiology. Genetic engineering permits us to probe deeply into the structure and, ultimately, function of many molecules essential to neural activity. Gene libraries, sequence analysis, the polymerase chain reaction, knock-out genetics and site-directed mutagenesis are among the powerful techniques revolutionising our picture of the fundamental events underlying brain activity.

5.1 Give an example of each of the following aspects of the structure and activity of informational molecules:
(a) molecular recognition
(b) transfer of sequence information
(c) cellular multiplication of molecular structures
(d) expression of information contained in a nucleotide sequence.

5.2 What is the value in knowing the base sequence of a particular gene? What information can we infer if the gene codes for a membrane protein?

5.3 A stretch of DNA consisting of two million base pairs is exposed for a long time to a restriction endonuclease. About how many fragments do you expect if the enzyme is (a) EcoR1 and (b) Hha1?

5.4 Compare a restriction enzyme that results in 'sticky ends' with one that yields 'blunt ends'. At what relative positions on the two strands does the enzyme cut in the two cases?

5.5 By what method can fragments of DNA be sorted by size? Which fragments move faster, small or large? How are they made visible?

5.6 A DNA molecule is digested by endonuclease A. One of the fragments, of length (in kbp) 4.3, when subjected to enzyme B, divides into fragments of lengths 2.8 and 1.5. When the DNA molecule is digested by enzyme B, one of the resulting fragments has length 3.9, which, when exposed to enzyme A, break into fragments of lengths 1.1 and 2.8. What is the likely relation between these fragments on a restriction map?

5.7 Describe the technique of homopolymer tailing. What enzymes and other reactants are used?

5.8 What is a plasmid? How does plasmid DNA differ from chromosomal DNA? How are plasmids used in gene cloning?

5.9 How does a bacteriophage act as a cloning vector? What are the characteristics of a γ-phage? What is the major advantage of a phage over a plasmid?

5.10 Is γ-phage DNA linear or circular in the capsid? In the bacterium? Describe the lytic stage of the γ-phage life cycle.

5.11 What are the *cos* sites of γ-phage? What enzymes recognise them? How does the genetic engineer use these structures to clone eukaryotic DNA by the cosmid technique?

5.12 What type of gene is incorporated into the plasmid vector to make possible the separation of bacteria with recombinant plasmids from the general population of bacteria? Describe the procedure.

5.13 What is a genomic DNA library? How is it produced? What difficulty arises in attempting to retrieve a particular gene?

5.14 What information is needed to generate a probe for fishing for a gene in a genomic library? How is the probe made? How is it used to select the gene of interest?

5.15 How is reverse transcriptase used to construct a cDNA gene library? By what means can the need for S1 nuclease be circumvented?

5.16 You are given a sample of a specific tissue. Describe a procedure by which you would generate cDNA and clone it.

5.17 Given a cloned cDNA plasmid, how would you use mRNA from the tissue to fish for a gene and express its protein product for physiological testing and identification?

5.18 Why is the amplification of a gene by cloning important for analysing the sequence of its DNA? Describe one sequencing technique.

5.19 What is the physical variable that is cycled in the polymerase chain reaction? Describe the PCR procedure for amplifying DNA. How is it modified when starting with mRNA?

5.20 Why is it sometimes said that the PCR 'extracts' a nucleotide sequence from the DNA double helix?

5.21 What is an expression vector? Give an example of a prokaryotic expression vector. What are its limitations?

5.22 Which large eukaryotic cell is frequently used as an expression vector? What are its advantages for manufacturing membrane proteins?

5.23 For a certain channel protein, it is desired to know the role of particular charged residue in the channel's conformational transition. The question arises as to what would be the effect of replacing this residue with a neutral one. What technique is called for? Outline the steps of a procedure leading to the expression of the alternate structure.

5.24 Explain how cultured neurons may be 'immortalised'.

5.25 Distinguish between 'heterologous' and 'homologous' recombination.

5.26 Describe the step by step procedure through which a gene may be 'knocked out' of a mouse genome. What human neurological condition has been studied using this technique?

5.27 What kinds of investigations are done by hybridisation histochemistry? Why is this technique particularly relevant to the nervous system? Describe the procedure and list a few of the available markers.

6

BIOMEMBRANES

In a sense, a membrane can be seen as a 'board' on which many of the 'circuit elements' of a cell are mounted. But this board, far from being rigid, is fluid, allowing for movement of embedded proteins. The phospholipid bilayer scaffolding is, moreover, an active component in its own right. Plasma membranes are continually synthesised and recycled. Multiple layers of glial or Schwann-cell membranes provide the myelin that affects the rate at which impulses are conducted. Interaction of the cell membrane with the cytoskeleton helps shape the cell, and junctions determine interactions between cells.

6.1 What are the principal features of the structure of a phosphoglyceride such as phosphatidylcholine?

6.2 Show how phospholipids aggregate to form bilayers, micelles and liposomes in aqueous media. What makes these structures stable?

6.3 Describe the motion of phospholipid molecules in a bilayer. What type of motion is generally not permitted?

6.4 What are the main features of glycolipid structure? What is a ganglioside?

6.5 With examples, describe the way in which gangliosides are named. What metabolic defect causes Tay–Sachs disease?

6.6 How is cholesterol arranged in a phospholipid bilayer? How does it affect membrane fluidity?

6.7 Describe the freeze-fracture technique. How does it help in determining the position of proteins in a membrane? Which is the 'P'-face and which the 'E'-face?

6.8 From which face of the lipid bilayer do the carbohydrate portions of the glycolipids project? In which 'leaflet' or 'monolayer' are the negatively charged phospholipids predominantly found?

6.9 Describe the fluid-mosaic model of a biomembrane.

6.10 Where would you expect to find the hydrophilic regions of a transmembrane protein? Where would you expect to find the hydrophobic regions? Explain.

6.11 Where do replacement proteins and lipids come from to repair an injured plasma membrane? What happens to the damaged parts of the membrane?

6.12 What types of cell form myelin in the central nervous system? What specific components are found in this myelin that are different from the plasma membrane of the parent cell?

6.13 How does the formation of central myelin differ from the formation of peripheral myelin? Can you suggest any reason for this difference? In which group of chordates does this mode of central myelination not occur?

6.14 With the help of a diagram, describe the position of the major proteins in peripheral myelin. To which family of proteins does protein zero (P_0) belong? What role does MBP play in holding the myelin whorl together?

6.15 Is proteolipid protein (PLP) found in peripheral or central myelin? What human neuropathology is caused by defective PLP?

6.16 Compare and contrast the cytoskeleton of neurons with that of erythrocytes.

6.17 What is the blood–brain barrier (BBB)? What is its neurological significance? From which protein structures is it formed? How do they function?

6.18 What structures function as electrical synapses? Draw a diagram, labelling the protein components.

6.19 Distinguish between connexins and connexons.

6.20 Are gap junctions more prevalent in embryonic than in adult nervous systems? In which part of an adult nervous system are they well represented?

6.21 Name a biochemical agent which uncouples gap junctions.

6.22 Distinguish, with examples, between 'metabotropic' and 'ionotropic' receptors.

6.23 Discuss the causation of the two major types of Charcot–Marie–Tooth (CMT) disease.

7

G-PROTEIN-COUPLED RECEPTORS

All cells possess devices for receiving molecular messages from other cells, both adjacent and distant. In this chapter, one such set of devices is examined: receptors, G-proteins, effector molecules and second messengers. An incoming signal molecule finds a receptor and binds to it, thus changing its conformation. In this new conformation the receptor is able to activate a G-protein. The activated G-protein splits, and a fragment migrates along the membrane until it collides with and thus activates an effector molecule. The activated effector catalyses the production of a second messenger, which diffuses into the cytosol. The whole complex system provides a means of amplifying the signal and inducing numerous subtle biochemical events in the receptor cell.

7.1 Distinguish between primary and secondary messengers.

7.2 What are the two possible outcomes of the formation of a ligand–receptor complex?

7.3 Name at least five members of the 7TM superfamily of receptors.

7.4 How and why are receptors desensitised? Distinguish between heterologous and homologous desensitisation.

7.5 What type of enzyme phosphorylates a protein? What type dephosphorylates it? Which amino-acid residues are phosphorylated, and at which chemical group?

7.6 G-proteins have been likened to 'precisely engineered time switches'. Explain the biochemical basis of this engineering analogy.

7.7 Which groups of regulatory proteins control G-protein time switches? In what way are receptors and effectors analogous to these regulatory proteins?

7.8 Explain the relation of membrane signalling G-proteins to the mammalian *c-ras* oncogene.

7.9 What second messenger does adenylyl cyclase induce? Distinguish between type 1 and type 2 adenylyl cyclase.

7.10 Which second messengers are induced by phospholipase C?

7.11 What major role does inositol triphosphate (IP_3 or $InsP_3$) play in a cell? (Look also at Box 9.2.)

7.12 Draw a diagram to show some of the many interactions between G-protein subunits in a membrane. How is inappropriate 'cross-talk' between these subunits avoided?

7.13 Describe the methodology used to isolate the β_2-AR gene. How were the seven membrane-spanning helices identified?

7.14 What event changes the conformation of a β_2-adrenergic receptor? What consequences follow from that conformational change?

7.15 What are the biochemical advantages of the G-protein system activated by noradrenaline?

7.16 How is the β_2-AR desensitised? What is the role of β-arrestin?

7.17 Distinguish between the regulatory and catalytic subunits of PKA.

7.18 What are the two types of receptors that are activated by acetylcholine? Contrast their distribution in the nervous system.

7.19 In what way does the neuroactivity of glutamate resemble that of acetylcholine?

7.20 Why is it believed that mGluRs have evolved independently of the other 7TM receptors?

7.21 How has the relation of structure to function been investigated in the NKRs?

7.22 How does retinal resemble the ligand of other 7TM receptors?

7.23 What is the initial event in photoreception?

7.24 What determines the wavelength to which a visual pigment is maximally sensitive?

7.25 What effector molecule is associated with opsin? What biochemical reaction does it catalyse?

7.26 How is opsin deactivated?

7.27 Which amino acid substitutions affect the absorption maxima (λ_{max}) of cone opsins?

7.28 Compare and contrast the structures of the four receptors: β-AR, mAChR, NKR and the opsins. What major features do they all have in common? How does the M1 muscarinic receptor differ from the others? The NKR?

8

PUMPS

The potential energy stored in concentration differences across cell membranes must ultimately come from metabolism. The conversion of metabolic energy into concentration differences is the function of pumps embedded in the plasma membrane. One glycoprotein, the sodium/potassium pump, converts the energy derived from the hydrolysis of one molecule of ATP to ADP to translocate three sodium ions outward and two potassium ions inward.

8.1 What is the function of a membrane pump?

8.2 What is a chemical potential? Under what condition will there be no net movement of a molecular species from one side of a permeable membrane to the other?

8.3 Use the equation $\Delta G = RT \ln (C_1/C_2)$ to calculate the energy required to pump one potassium ion into a squid axon.

8.4 After an $Na^+ + K^+$ pump has been in operation for some time, an experimenter determines that 1800 potassium ions have been pumped into the cell. How many sodium ions have been pumped out? How many molecules of ATP were required? What is this mechanism called and what is its efficiency?

8.5 Name two cardiac glycosides that inhibit the $Na^+ + K^+$ pump. Where on the molecule do they act?

8.6 Where on an $Na^+ + K^+$ pump is the carbohydrate located? Describe the molecule. Where does the catalytic action take place? Where does ouabain bind?

8.7 What happens to the phosphate group released from ATP in the hydrolysis? What does the energy release do to the α-subunit?

8.8 What type of evidence suggests that the calcium pump of red blood cells differs from that of striated muscle fibres?

8.9 What is the function of sarcoplasmic reticulum? What is the function of the Ca^{2+} pump in the SR membrane?

8.10 Which other ion plays a role in Ca^{2+} pump function? What is that role believed to be?

8.11 What parameter regulates the working rate of the Ca^{2+} pump? What protein is involved in this regulation?

8.12 Explain the binding of Ca^{2+} ions to calmodulin. Outline how the binding of Ca^{2+} 'activates' calmodulin so that it can interact with other cellular molecules and structures.

8.13 How do the molecular structures of the $Na^+ + K^+$ and the Ca^{2+} pumps resemble each other?

8.14 What other pumps and transport mechanisms play important roles in neural membranes?

9

LIGAND-GATED CHANNELS

Ligand-gated ion channels are membrane proteins that respond with a conformational change to the binding of one or more molecules of a specific substance. In the activated state these channels are highly selective conductors of ions, which pass across the membrane as an electric current. Ligand-gated ion channels are usually located in the subsynaptic membrane of a synapse. They are named after the substances that, by binding, activate them, such as acetylcholine, γ-aminobutyric acid (GABA), glycine etc. The unfolding story of the great variety of excitatory and inhibitory receptors of differing characteristics is adding new dimensions to our conception of brain function.

9.1 What are the differences and the similarities between ligand-controlled and voltage-controlled channels?

9.2 What are the two types of cholinergic receptor? How do they differ in response to acetylcholine?

9.3 Which areas of the brain are richest in central nicotinic acetylcholine receptors? Where would you expect to find peripheral nAChRs?

9.4 How do you account for the great similarity between the electric organs of *Torpedo* and *Electrophorus*? Are the electric rays and eels closely related?

9.5 What role do neurotoxic peptides play in isolating nAChR? Outline a procedure for determining the molecular weights of the receptor's subunits.

9.6 Explain how, given a small quantity of purified receptor, you would determine its amino acid sequence. How would you verify that the amino acid sequence you have determined is indeed that of the nAChR?

9.7 Describe the subunit structure of nAChR from *Torpedo* electroplax. Comment on the differences and similarities with nAChR from other sources. How many subunits constitute the complete receptor?

9.8 What information in the primary structure indicates the parts of the subunit that probably pass through the lipid bilayer? How many such stretches are found in *Torpedo* nAChR subunit?

9.9 How do the electron microscope data on nAChR confirm the data from sequence analysis? What additional information do the electron micrographs provide?

9.10 Describe the liposome technique for investigating nAChRs. What results does its use yield?

9.11 How does a patch-clamper obtain an inside-out patch, an outside-out patch and a whole-cell clamp?

9.12 How does the number of univalent ions passing through a single ion channel during a discrete opening event depend on the single-channel current and on the duration of channel opening?

9.13 How many ACh molecules are required to open an nAChR channel? Are the binding events independent? Explain. (Recall the way oxygen molecules bind to haemoglobin.)

9.14 What happens if ACh molecules remain in contact with the receptor for a period greater than about ten seconds? What biochemical mechanism is believed to be at work?

9.15 List the major events in the synthesis and assembly of an nAChR in a skeletal muscle fibre. Where are the subunits assembled? Where does the mature receptor form? To what proteins are the subunits anchored and how?

9.16 Explain what is meant by the term 'half-life'. What is the half-life of the nAChR in the postsynaptic membrane?

9.17 What is the effect of innervation of a striated muscle on the distribution of nAChRs? What protein is secreted to bring about this effect?

9.18 What change in mammalian nAChR response occurs when the neuromuscular junction is established? What is the genetic and structural basis for this change? (Compare the development of haemoglobin.)

9.19 Compare and contrast myasthenia gravis and Lambert–Easton myasthenia (LEM).

9.20 What physiological and pharmacological differences distinguish central (i.e. neuronal) nAChRs from peripheral (i.e. muscle) nAChRs? How do the subunit structures of the two classes of nAChR differ?

9.21 How would you estimate the evolutionary relationships of nAChRs? Why is it believed that neuronal nAChRs are the more ancient?

9.22 List two subtypes of GABA receptors. Where in a synapse is each subtype located? What is the mode of action of each?

9.23 How many subunits does $GABA_A$ have? How many of the subunits are different? How many hydrophobic stretches are found in a subunit? What is the structural interpretation of these stretches?

9.24 How are glycine receptors isolated? Discuss their polypeptide structure.

9.25 What is the procedure for clarifying the positions of the subunits of a polypeptide on a membrane by immunoelectron microscopy? What are the results for GlyR?

9.26 How many glutamate receptors have been isolated?

9.27 List the ionotropic glutamate receptors and indicate the pharmacological agents used to discriminate between them.

9.28 How many polypeptides comprise the AMPA and KA receptors? How many transmembrane segments are there in each polypeptide? What ions pass through the channel?

9.29 What role does mRNA editing play in the synthesis of some of the AMPA and KA receptors?

9.30 In addition to its ligand sensitivity, the NMDA receptor is also sensitive to a physical variable and affected by a divalent ion. What variable? What ion? Describe the interaction between them.

9.31 Discuss some of the effects that make the NMDA receptor a candidate molecule for associative learning and short-term memory.

9.32 Where are IP_3 receptors located? Which type of ion channel do they control? Why are they important in the life of the cell?

9.33 How has the ligand-binding site of the IP_3 receptor been determined?

9.34 What are the agonists of the purinoceptors? Discuss the structure of these receptors.

9.35 Construct a diagram to show the interrelatedness of ligand-gated ion channel.

10

VOLTAGE-GATED CHANNELS

> Voltage-dependent ion channels sit squarely at the intersection of physics and biology. As physical devices, they transduce small changes in transmembrane voltage into large changes in ion current. As biological molecules, they are a set of evolutionarily related glycoproteins adapted to a great variety of functions.

10.1 What are the major types of voltage-gated channels? What are their functions?

10.2 Describe the conventions used for naming voltage-gated channels. What is meant by the symbols $K_{5\text{-HT}}$ and $I_{K(5\text{-HT})}$?

10.3 What is the magnitude of a typical resting potential? How thick is a typical plasma membrane? Use these figures to calculate the average electric field (potential gradient) across the membrane.

10.4 Name two substances that immobilise the Na^+ channel. How many of these molecules are required to deactivate one Na^+ channel?

10.5 How do sodium channels in mammalian brain and skeletal muscle differ from the channels found in eel electroplax? What explanation has been proposed to account for the increased complexity?

10.6 Explain how a neurotoxin can be used as a probe molecule to isolate and purify Na^+ channels from a membrane.

10.7 How many homologous domains does the α-subunit of the Na^+ channel have? What does the homology suggest as to the evolution of the channel?

10.8 Which membrane-spanning segment of each Na^+ channel domain is considered to be the 'voltage sensor'? What structural property of this segment suggests this concept?

10.9 A patch of nerve membrane about 0.5 μm in diameter is voltage-clamped. Sketch a possible current response when the membrane is depolarised by 10 mV for 15 ms. How is the total current explained in terms of single-channel currents?

10.10 List and describe the three major conformations of the Na^+ channel. At what phases of the channel's activity cycle do they occur?

10.11 A single channel passes 2 pA during an opening of 0.8 ms. How many ions traverse the membrane? ($e = 1.6 \times 10^{-19}$ C.)

10.12 How can we measure the number of Na^+ channels per area of membrane? Where would you expect to find the highest density of channels: at the node or in the internodal region of a myelinated fibre, or in an unmyelinated fibre? Explain.

10.13 Is there a specific marker for K^+ channels? If so, what is it? If not, what special methods permit the molecular study of certain K^+ channels?

10.14 Name four major types of potassium channels and describe their principal characteristics.

10.15 What is the shaker mutant? What is the electrophysiological appearance of the defect, and where does it occur? What cell is used for the expression of this polypeptide?

10.16 Which pharmacological agents inactivate the fast K^+ channel? Which part of the molecule is responsible for this inactivation and how has this been shown?

10.17 How can a variety of differing K^+ channels be derived from a single *shaker* gene?

10.18 How does the delayed K^+channel (K_V) differ from the fast K^+channel (K_A)?

10.19 Explain what is meant by 'anomalous rectification'. Where were GIRK channels first discovered and what physiological role do they play? (See also Section 7.9.)

10.20 How does the molecular structure of GIRK channel subunits resemble that of K_{ATP} subunits?

10.21 Where on a neuron would you expect to find Ca^{2+} channels? What important roles do they play?

10.22 What is the approximate internal free calcium ion concentration of a neuron? Can the opening of Ca^{2+} channels affect this concentration significantly? Explain.

10.23 What channel does the Ca^{2+} channel resemble closely? What characteristics do these channels have in common? How do they differ?

10.24 How does the S4 segment of the Ca^{2+} channel resemble the S4 segment of the K^+ and Na^+ channel subunits? What is believed to be the function of this segment?

10.25 Explain the sliding helix conjecture of the voltage sensor and contrast this with the 'ferroelectric' conjecture.

10.26 Which part of the subunit structure of voltage-sensitive channels specifies the ion selectivity?

10.27 List the Cl^- channels which have been isolated to date. Which is the smallest channel subunit of all?

10.28 Give an account of the evolution of ion channels. Which channels are believed to be ancestral and why? Why is it believed that all of the ligand-gated ion channels (LGICs) and the voltage-gated ion channels (VGICs) form separate though evolutionarily related families?

10.29 With reference to Figure 10.21, review the molecular architecture of G-coupled receptors, LGICs and VGICs.

11

RESTING POTENTIALS AND CABLE CONDUCTION

A neuron's 'vital sign' is its resting potential, which provides the necessary condition for its ion channels to function. The resting potential originates in passive diffusion of ions across the cell membrane, mainly the outflow of potassium ions. Carrying their positive charges out, they leave negative intracellular charge behind. The resulting electric field resists the efflux of the potassium ions. The opposition of the two effects, diffusion and migration, results in the equilibrium condition that provides the stability of the resting potential. Knowing the concentrations of the diffusible ions on each side of the membrane, this potential can be calculated by means of the Goldman equation.

11.1 Which animals have giant axons? In what ways have these axons been useful to electrophysiologists?

11.2 A membrane separates two phases at different electrical potentials and with different ion concentrations. What is the condition for equilibrium between the two phases? What is the name of the parameter used?

11.3 The Nernst equation is $V = (RT/Z_I F) \ln (C_0/C_i)$. Explain the meaning of V, R, T, Z_i, F, C_0 and C_i.

11.4 Using the values for Na^+ given in Table 8.1, work out the values of V_{Na} for squid giant axon and cat motor neuron. Why does this value differ so markedly from the resting value of V_m?

11.5 In the Goldman equation, what are the *P*s called? Which ion has the biggest effect on the resting potential in neurons and glia?

11.6 Insert $P_{Na} = 5 \times 10^{-7}$ cm/s into the Goldman equation, keeping all the other permeabilities unchanged and using the ionic concentrations of Table 8.1. Calculate the value of V_M. Refer to page 263 of the text and compare your value with that observed in rod outer segments in the dark.

11.7 What are Müller cells and where are they found? How does their permeability vary from that of neurons? What physiological role does this confer on these cells?

11.8 Write down the simplified version of the Goldman equation. What does the symbol b stand for?

11.9 What is the name given to small changes in membrane potential that do not initiate action potentials? How do the currents caused by them distribute themselves in space?

11.10 In what way is an axon like an undersea cable? Other than in size, how does it differ?

11.11 In the equation $V = \lambda^2 d^2V/dx^2$ what is λ called? What is its significance?

11.12 What is the effect on λ of an increase in the diameter of a process?

11.13 What is the charging time constant of a neuron? How does it change as the stimulus is brought closer to the perikaryon along the dendritic tree?

12

SENSORY TRANSDUCTION

All organisms are sensitive to changes in their environment. Sense organs in 'higher' animals such as ourselves are often hugely complex. The molecular mechanisms of our sensory cells nevertheless often retain traces of their prokaryotic origins. Cilia in various guises play key roles in the detection of odours, tastes, light and sound.

12.1 Does a nerve impulse carrying smell information differ from one relaying sight or sound? If so, how? If not, what allows the impulse to carry its proper message to the brain?

12.2 What is a receptor potential? When does it become a generator potential? In what type of cell are the two types of functions combined?

12.3 Describe the two types of flagellar rotation that occur in bacteria. How does each type affect the motion of the cell? How does the cell's location in a gradient of chemical attractant change the migration pattern?

12.4 Which attractants act directly on the transmembrane proteins? By what mechanism do other attractants affect the transmembrane proteins? Which attractants?

12.5 Describe sensory adaptation in bacteria. What is another name for this phenomenon? Does adaptation to one attractant generalise to other molecules? Explain.

12.6 Which receptor–transducer protein responds to aspartate? What gene codes for it? How does this R–T molecule respond to nickel ions?

12.7 How many membrane-spanning segments does an R–T polypeptide have? What is the role of the cytoplasmic domain? What is the role of the periplasmic domain? How many polypeptides constitute a complete R–T protein?

12.8 Which proteins cause the flagellum to rotate clockwise? Does an attractant or a repellant stimulate release of one of these proteins? What bacterial motion results?

12.9 Describe the molecular mechanism sensory adaptation in *E. coli*. What is the biological significance of sensory adaptation?

12.10 Which types of cell make up the olfactory epithelium of mammals? Sketch the structure of the epithelium, showing the olfactory 'knob', cilia and other features.

12.11 What organelle that commonly provides motility is adapted to serve sensation? Describe its structure. What types of molecules does its membrane contain?

12.12 What roles in olfactory reception do the following structures play?
(a) 7TM receptor
(b) G-protein
(c) adenylyl cyclase
(d) cAMP
(e) CNG channel.

12.13 How does the binding of a cAMP to a CNG channel lead to a receptor/generator potential? In what way does the information reach the brain?

12.14 Outline the major events of gustatory reception. How does it differ from olfactory reception?

12.15 Sketch a cone cell, labelling the major components.

12.16 What lines of evidence show that the outer segments of rod and cone cells evolved from cilia?

12.17 Which membrane of a rod cell contains rhodopsin molecules? Which membrane contains ion channels? What ions pass through these? When are they open?

12.18 Using the Goldman equation and ion concentrations given for cat motor neuron in Table 8.1, show that, when the Na^+ and K^+ permeabilities are respectively five and ten times the Cl^- permeability, the equilibrium membrane potential difference is about $-20\,mV$.

12.19 What nucleotide keeps the CNG channels in the outer segment open? What is the source of energy for maintaining the dark current in rod and cone cells?

12.20 Compare and contrast the functioning of the CNG gates in outer-segment membranes with those in the membranes of olfactory cilia.

12.21 Does a flash of light produce a depolarisation or a hyperpolarisation in the synaptic foot of a rod cell? Does it increase or decrease transmitter release? What response is observed in the subsynaptic bipolar cells?

12.22 What is the role of recoverin in retinal adaptation to sustained illumination? How does it become activated?

12.23 What symptoms are shown by patients suffering from retinitis pigmentosa (RP)? Which chromosomes are believed to be involved?

12.24 What molecular pathology is responsible for some types of retinitis pigmentosa? Has animal experimentation given hope of possible therapy?

12.25 State two important roles mechanoreceptors play in the life of a cell.

12.26 How can prokaryote plasma membrane be patch-clamped? What mechanosensitive channels are found?

12.27 What are the degenerins? How have they been linked to the sense of touch?

12.28 What roles do hair cells play in vertebrates?

12.29 Name the compartments of the membranous labyrinth and specify their functions. What fluids does it separate? How do they differ in ionic concentration?

12.30 Sketch a hair cell of the membranous labyrinth, labelling the movable parts, synapses, adjacent cells etc.

12.31 What are the electrical effects of moving a stereocilium towards the kinocilium, away from it and at right angles to this axis?

12.32 Which ions flow into the stereocilia in response to movement? Why is it thought that these ions affect the membrane potential directly without the help of any intervening G-protein biochemistry?

12.33 What are microphonic potentials? How are they related to the movement of stereocilia?

12.34 Why is it impossible for auditory nerve fibres to respond to high frequency microphonic potentials with a matching high frequency of action potentials?

13

THE ACTION POTENTIAL

The impulse that is propagated along the nerve fibre is detectable as an electric pulse, the action potential. Starting from its springboard, the negative resting potential, the voltage trace on an oscilloscope screen swiftly leaps up into the positive region of the sodium reversal potential only to fall more slowly back to the resting potential. Underlying and responsible for this simple picture are the conformational changes of vast numbers of ion channels: first the sodium channels open for the inward current, then after a delay the potassium channels open for the outward current.

13.1 Sketch an action potential. Indicate voltage and time scales.

13.2 What is a voltage-clamp circuit? What is it used for?

13.3 Describe the current as a function of time recorded from a voltage-clamped axon that has been (a) depolarised to −5 mV and (b) hyperpolarised to −125 mV.

13.4 Describe the current from an axon that is immersed in a solution in which Na^+ has been replaced by an impermeant cation and that is clamped to −5 mV. What ion carries this current?

13.5 Consider the difference obtained by subtracting the current in Question 13.4 from that in 13.3a. Is this current inward or outward? Describe its shape. What ion carries it?

13.6 (a) If the capacitance of a membrane is 1.2 μF/cm^2, what is the charge on one side of a patch of membrane of area 1 μm^2 when the voltage is 80 mV? (b) How many ions is that? ($e = 1.6 \times 10^{-19}$ C.)

13.7 How does one obtain (a) an inside-out patch and (b) an outside-out patch?

13.8 Do Na^+ channels remain open as long as a depolarising voltage is applied? If not, what occurs?

13.9 Explain how the increase in Na^+ conductance with depolarisation leads to a positive feedback loop.

13.10 The actual sodium current measured in a patch-clamp experiment is less than that calculated from the open-channel conductance and the channel density. How can that be explained?

13.11 How do local circuits explain the spatial propagation of the action potential?

13.12 If an axon is stimulated symmetrically by an electrode, in which direction, or both, will the impulse be propagated? In which direction does an impulse propagate *in vivo*? Why?

13.13 How does the inactivation of the Na^+ channel lead to a refractory region in the membrane?

13.14 If inactivation lasts for 1 ms, how wide is the refractory region in an axon with propagation velocity 5 m/s?

13.15 Why do the ion gradients not run down after the passage of many action potentials along a living neuron?

13.16 How do invertebrates achieve high propagation velocities? How do vertebrates achieve them?

13.17 How does myelination affect λ, the space constant or electrotonic length? Why is this effect important and what happens when demyelination occurs, as in cases of multiple sclerosis (MS)?

13.18 Make a sketch showing the relation between an axon and the glial cell investing it in its myelin sheath. Label nodes, internodes and paranodes. What are the functions of the paranodes?

13.19 Explain what happens in saltatory conduction.

14

THE NEURON AS A SECRETORY CELL

Study of the neuron as a secretory cell is very active at the time of writing. Many of its characteristics in this regard are specialisations and developments of the molecular biology of non-neural secretory cells. The student would do well to review material in textbooks of general and molecular biology, for example the standard text by Alberts *et al.*, 1994.

14.1 State Dale's 'law.' Is it always true? If so, what are some of its implications? If not, state, an exception.

14.2 Two nearby terminals are found to secrete different neurotransmitters. Is it likely that they originate in the same neuron, or different neurons? Why? Can you be certain, without further evidence?

14.3 Criticise the analogy of an axon to a telegraph wire. Can you think of an analogy to the axon's 'non-telegraphic' function? Hint: Along what structures are telegraph (or telephone) wires frequently situated?

14.4 There are many different neurotransmitters (see also Chapter 15). Where are they synthesised?

14.5 Draw a labelled diagram of the perikaryon of a typical multipolar neuron. Write notes on the functional significance of the organelles you have labelled.

14.6 Trace the flow of information from the genes to the neurotransmitters and the enzymes needed to produce them.

14.7 Explain the processes by which proteins enter the lumen of the endoplasmic reticulum. Many neurobiologically important proteins are membrane-bound. Give some examples. Discuss how they achieve their position.

14.8 Where on the ER, in the lumen or outside, would you expect to find the amino terminal of a membrane-spanning protein? Where would you expect to find the carboxy terminal if the polypeptide passes through the membrane an odd number of times? An even number of times?

14.9 Where in the cell are disulphide bonds formed? What is the function of these bonds?

14.10 Describe the nature of the Golgi apparatus and explain its role in the biology of the cell. With the aid of appropriate chemical formulae, distinguish between N and O glycosylation.

14.11 What are the functions of transport and secretory vesicles? Describe their structural design elements.

14.12 How would you demonstrate the existence of flows of material within an axon? How would you measure the velocities of these flows?

14.13 Consider an axon 1 m long. How long does it take (give a range of times) for a secretory vesicle to work its way from soma to terminal? How long for cytoskeletal actin? How long for the return trip of materials being recycled?

14.14 Construct a diagram to show the major elements of the cytoskeleton within an axon. Explain how these elements are involved in the production of axoplasmic flow.

14.15 What methods would you use to demonstrate the existence of different fibrous elements within an axon and their relative distribution?

14.16 Compare microfilaments with intermediate filaments on the basis of:
(a) diameter and length
(b) type, shape and size of monomer
(c) nature of polymerisation
(d) cross-linking behaviour.

14.17 Describe the structure of microtubules. Are they axially symmetrical or polarised? Explain treadmilling.

14.18 What is AVEC microscopy? How has it been used to study axonal transport?

14.19 An axon from a basal nucleus to the cerebral cortex may be about 10 cm in length. Calculate the time take for a neuropeptide synthesised in the perikaryon to reach the synaptic terminal.

14.20 How do cytoplasmic dynein and kinesin transport material in axons? How does the mechanism differ from that adopted by dynamin?

14.21 Relate the size and appearance of a synaptic vesicle to its contents. What type of vesicles carry excitatory neurotransmitters? Which carry inhibitory ones?

14.22 Are all the contents of a synaptic vesicle necessary for its job of transmitting information across the synaptic cleft, or do some appear to be 'excess baggage'?

14.23 The neuromuscular junction has been called the *E. coli* of the nervous system. Discuss. (See also Section 18.9.)

14.24 Compare the parts played by synapsin 1 in synaptic terminals with the roles of band 4.1 protein in erythrocytes.

14.25 What is the role of Ca^{2+} in the release of synaptic vesicles? Where are the Ca^{2+} channels located? How do we know?

14.26 Construct a table showing the major molecules involved in vesicle docking. Indicate the part which each is believed to play.

14.27 Compare the ultrastructure of presynaptic membranes at the neuromuscular junction and at central synapses. What use has been made of aqueorin and ω-conotoxin in analysing the functional significance of this ultrastructure?

14.28 What are the functions of:
(a) coated pits
(b) multivesicular bodies
(c) calmodulin?

14.29 Referring back to Chapter 7 and Box 7.1, explain how rab3A is believed to control the docking of vesicles on the P-face of the presynaptic membrane.

14.30 Provide an illustrated account of the recycling of vesicle membrane at the presynaptic terminal. Why is this recycling necessary?

14.31 Describe the energetics of vesicle refilling in the synaptic terminal.

14.32 How has the vesicular acetylcholine transporter (VAChT) been isolated? Why is the genetic structure of the VAChT and ChAT genes an economical arrangement?

15

NEUROTRANSMITTERS AND NEUROMODULATORS

This chapter covers in short material to which texts in neuropharmacology give extended treatment. The student wishing further information should consult one of the many excellent texts in the latter subject. When perusing this chapter the student should also bear in mind the earlier chapters of this book, which give extensive treatment of the various receptors and ion channels that the transmitters and modulators affect.

15.1 At what two locations in a neuron are neurotransmitters and neuromodulators synthesised? Where are the enzymes necessary for their synthesis produced? How do they get to the locations where they are needed?

15.2 Criticise the telephone-exchange metaphor of the central nervous system. In what sense is it valid? In what ways does it fail?

15.3 With examples, explain the difference between neurotransmitters and neuromodulators.

15.4 Cite an example of a neurotransmitter that works by either ionotropic or metabotropic action. Discuss the locations, time scales and mechanisms of these two actions.

15.5 Give an example each of an indoleamine, a catecholamine, an opioid peptide and a tachykinin.

15.6 List the criteria that should be satisfied if a candidate molecule is to be confirmed as neurotransmitter or modulator.

15.7 Review Sections 7.9 and 9.1 and write a short account of the differences between the muscarinic and nicotinic action of acetylcholine.

15.8 What lines of evidence suggest that cholinergic neurons are involved in our ability to remember?

15.9 Can receptors be found in the presynaptic membrane of a synapse as well as the postsynaptic membrane? Give an example. What function might such a placement have?

15.10 What role does ChAT (choline acetyl-transferase) play in cholinergic synapses?

15.11 How would you identify cholinergic synapses in a section of the brain?

15.12 What is the affect on a cholinergic synapse of the following pharmacological agents:
(a) atropine

(b) physostigmine (eserine)
(c) α-bungarotoxin
(d) β-bungarotoxin
(e) hemicholinium
(f) botulinum toxin
In each case indicate where the agent has its major site of action in the synapse.

15.13 What major neuropathology is believed to be associated with loss of central cholinergic fibres? (See also Section 20.9.)

15.14 Write down the chemical formulae of the major amino acid neurotransmitters. Do they have ionotropic or metabotropic action? (Check also Sections 7.10 and 9.4.) Which amino acids are excitatory and which are inhibitory?

15.15 Describe the function of glial cells in glutaminergic terminals.

15.16 How is the activity of amino acid synapses terminated? Does this means of termination differ from that found in cholinergic, serotoninergic and adrenergic synapses?

15.17 What is the function of glutamic acid decarboxylase? In what cells, and where in the cell, is it found? What coenzyme does it require?

15.18 In what ways do $GABA_A$ and $GABA_B$ receptors resemble nACh and mACh receptors?

15.19 Which ion-specific channel is controlled by the $GABA_A$ receptor? Show, by using the Goldman equation and the customary distribution of ions across neural membranes, which way the membrane potential would move if this channel were opened.

15.20 Which neurological condition is believed to be associated with loss of GABAergic synapses? Where are these synapses located and why should the loss of GABAergic synapses in this location lead to the symptoms of the disease? (See also Section 20.6.)

15.21 Which receptors does strychnine block? Why should blockade of these receptors lead to generalised convulsions and often death?

15.22 In which regions of the brain is serotonin (5-HT) most concentrated?

15.23 What are varicosities? Why should activation of serotoninergic neurons in the raphe nucleus lead to widespread cortical response?

15.24 List the classes and subclasses of serotonin receptors. Are all found in the brain? Explain. Which of the receptors are found in presynaptic membranes? What is their function?

15.25 How does Prozac (fluoxetine) exert its antidepressant action?

15.26 Name three pharmacological agents that deplete serotoninergic terminals of their transmitter. Explain the biochemistry that ensures this depletion.

15.27 Explain how dopaminergic and noradrenergic fibres can be localised by the appropriate administration of 6-hydoxydopamine (6-OHDA) and desipramine.

15.28 Where in the brain do catecholaminergic neurons have their cell bodies? In what parts of the brain do they principally terminate?

15.29 Draw the pathway by which norepinephrine is synthesised from phenylalanine, indicating the enzymes and co-factors involved at each step.

15.30 Which neurotransmitter is defective in Parkinsonism? What can be done to temporarily relieve the symptoms of the condition? (See Section 20.8 for a fuller discussion.)

15.31 Describe the 'push–pull' modulation of transmitter release from noradrenergic terminals.

15.32 How are neurotransmitters sequestered into vesicles? (Review Section 14.4 and Box 14.2.) How does reserpine affect this process?

15.33 Which enzymes inactivate catecholamines and what are the synaptically inactive products of their action? What happens to these inactive end-products?

15.34 How many transmembrane domains has the dopamine reuptake transporter?

15.35 Why is the dopamine reuptake transporter most strongly concentrated in the substantia nigra?

15.36 Reuptake transporters are classified as 'symporters', unlike vesicular transporters, which are examples of 'antiporters'. Explain the difference between the two classes of transporter.

15.37 Explain what is meant by the acronym NANC.

15.38 How would you discriminate between P_1 and P_2 purinoceptors?

15.39 How may purines be of use in the treatment of stroke?

15.40 Why are enkephalins (and dynorphin) called 'endogenous opioids'?

15.41 Where is the 'pain pathway' in the CNS? What transmitters does it use and how is its activity controlled? What biological purpose is served by controlling this pathway?

15.42 What is denervation sensitivity and how may it account for the painful withdrawal symptoms experienced during 'drying out' after substance abuse?

15.43 How does a neurosecretory cell differ from a peptidergic neuron? Describe the linkage between peptidergic and neurosecretory cells in the hypothalamo-pituitary system.

15.44 Discuss the neurophysiological significance of cohabitation (or co-existence) of peptide and non-peptide transmitters in a single neuron.

15.45 How does nitric oxide (NO) differ from the other neurotransmitters and neuromodulators considered in Chapter 15?

15.46 What enzyme is required to synthesise NO? Some forms of this enzyme are said to be 'inducible', others 'constitutive'. What is the difference?

15.47 Does NO have specific receptors in neural membranes, or does it merely diffuse into the cytosol of neurons?

15.48 What is the principal target molecule for NO? When NO interacts with this target what molecule is synthesised?

15.49 Compare and contrast the 'hardware' (better, 'wetware'!) of the brain with that of electronic 'brains'.

15.50 What is the biogenic amine theory of depression? Cite several lines of evidence supporting it.

15.51 Describe the biochemical pathways by which unsequestered dopamine is rendered synaptically inactive.

16

THE SUBSYNAPTIC CELL

This chapter brings together many of the elements discussed in earlier chapters. It synthesises the material on membranes, receptors, ion channels, G-protein systems, first and second messengers, IEGs and third messengers dealt with in detail in earlier chapters. In studying this chapter, students should refer to those earlier chapters and boxes.

16.1 What are synaptosomes and how may they be prepared?

16.2 What is the synaptic plasma membrane (SPM)?

16.3 What is the postsynaptic density (PSD)?

16.4 What proteins may be isolated from PSD? What roles do they play in the operation of the synapse?

16.5 What is the meaning of the acronyms EPSP and IPSP?

16.6 Describe the ionic flow responsible for an EPSP. Show how an EPSP can lead to the initiation of an action potential.

16.7 Suppose the permeability constant for potassium is 100 times that for chloride, and that for sodium is 75 times that for chloride. If the ion concentrations are those for cat motor neuron, what membrane voltage does the Goldman equation predict?

16.8 Suppose the relative permeability constants for potassium, sodium and chloride are in the proportions 10:1:100, respectively. If the ion concentrations are those for cat motor neuron, what membrane voltage does the Goldman equation predict?

16.9 How does the activation of GABA or glycine receptors lead to a membrane hyperpolarisation? Compare with what happens at an excitatory synapse.

16.10 How does the magnitude and location of a synapse affect its influence on a subsynaptic cell?

16.11 What is meant by temporal and spatial summation and how are these phenomena important in the physiology of the CNS?

16.12 Explain what is meant by 'algebraic summation' of EPSPs and IPSPs at an axon's initial segment.

16.13 Define an ion's 'reversal potential'. If $V_K = -85\,mV$ and $V_m = -55\,mV$ calculate the force moving a K^+ ion across the membrane.

16.14 Is the opening of a single ion channel in the subsynaptic membrane a simple 'all-or-nothing' event? Discuss, with examples.

16.15 Apply Equation 16.1 to the acetylcholine receptor. Let *T* stand for ACh and *R* for AChR. Explain what is meant by *TR*.

16.16 Compare the two-state model of ion channels with data showing multiple conductance states in receptors.

16.17 If taurine is known to open channels and its mean channel opening time has been measured, why is it considered a 'doubtful' neurotransmitter? Hint: Consult Box 15.1.

16.18 How does the response of a population of ion channels to stimulation differ in the subsynaptic membrane from that found at the axon's initial segment or in the node of Ranvier? Explain the cause of this difference and indicate why the difference is significant in neurophysiology.

16.19 Consider an axo-axonic synapse, in which the distal bouton releases noradrenaline. How are the Ca^{2+} channels of the proximal bouton affected? How is the response to an action potential arriving at the proximal bouton changed? What is the result on the subsynaptic cell?

16.20 Review Section 7.5 and draw up a list of second messengers.

16.21 Review Section 7.5 and Box 9.2, and after reading Section 16.6 discuss the importance of Ca^{2+} as a second messenger.

16.22 Draw a diagram to illustrate the mechanism by which activation of muscarinic cholinergic affects M channels. How does this influence (a) the length of the action potential and (b) the firing rate of the neuron?

16.23 Review Sections 3.3.1, 3.4.2 and Box 3.1 and explain (with examples) what is meant by a 'third messenger'.

16.24 What are IEGs and how were their transcripts first detected in the CNS?

16.25 Draw a diagram to show how dopamine induces the synthesis of enkephalins by enkephalinergic neurons.

16.26 What is the meaning of the acronym CREB? How is the CREB site important in some peptide-ergic cells?

16.27 How are Ca^{2+} ions involved in initiating the transcription of neuropeptide genes?

16.28 Where is the pineal gland located? What hormone does it secrete? What is the significance of this hormone in many infra-human animals? What is the role of the hormone in humans?

16.29 List the neurotransmitters that stimulate pinealocytes.

16.30 Are IEGs involved in the response of pinealocytes to noradrenaline?

16.31 What is the rate-limiting enzyme in the synthesis of melatonin from serotonin?

17

GENETICS OF THE BRAIN

The guiding principle in Chapter 17 is that control of transcription gives control of the differentiation of the cells in the CNS. The genes that encode transcription factors are deeply embedded in the evolutionary development of living organisms. They can be traced back, hardly changed, over a billion years. The cells that are fated to form the nervous system are at first generalised and pluripotent. Transcription factors determine which proteins are synthesised and hence which type of cell develops. As there are many more types of cell in the CNS than anywhere else in the body, neurogenetics reveals highly intricate cascades of genes and gene products. Analysis of these cascades is only just beginning. The student can follow the developing understanding in *Trends in Neurosciences*, *Trends in Genetics*, and review articles in *Neuron*, *Cell Developmental Biology*, *Genes and Development* etc. The student should also note the convention for writing the names of genes and proteins. It derives from an earlier convention for the expression of the names of species. These should always be italicised: *Homo sapiens*, *Mus musculus* etc. Similarly the names of genes and gene transcripts should be italicised; the names of proteins, in contrast, are expressed in roman script. If this convention is followed much confusion is averted (for further information see the introductory paragraph to Appendix 4).

17.1 Review Sections 1.2 and 17.1 and explain what is meant by the 'biogenic law'. Why should genetic damage affecting early development be more lethal than damage affecting later stages?

17.2 What is meant by the term 'heterochrony'?

17.3 At what stage in development is the anteroposterior (A-P) axis established in *Drosophila*?

17.4 What is a syncytial blastoderm and when is it transformed into a cellular blastoderm? Does a syncytial blastoderm develop in early vertebrate embryology?

17.5 Which gene establishes the anterior end of the early *Drosophila* embryo? What happens if this gene is inactivated?

17.6 What role could gap junctions play in establishing the A-P axis in a multicellular blastula? Is there any evidence that they play this role?

17.7 What is the function of the 'pair rule' genes in *Drosophila*?

17.8 Where and how do neural crest cells (NCCs) originate? What structures are they destined to form?

17.9 What is the sympatho-adrenal (SA) lineage and what use has been made of it by embryologists?

17.10 How is the A-P axis of the vertebrate neural tube established?

17.11 Draw a diagram showing the initial stages in the dorsoventral differentiation of the neural tube.

17.12 *Wnt-1* and *en-2* are genes controlling early segmentation in the mouse. What are their analogues in *Drosophila*? How may their functions be analysed in mammals?

17.13 What is the meaning of the term 'homeosis'?

17.14 What phenotypic changes are observed when *Drosophila* bithorax (*bx*) and antennapedia (*antp*) genes mutate?

17.15 What do the abbreviations *ANT-C* and *BX-C* stand for?

17.16 Define the terms 'homeobox' and 'homeodomain'.

17.17 What structural features of the homeodomain suggest that it functions as a transcription factor? (Review Section 3.3.1 and Figure 3.6.)

17.18 Differentiation proceeds by a cascade in which the transcript of one gene controls the transcription of another, and so on. Which genes control the transcription of homeobox genes in the early embryo of *Drosophila*?

17.19 What is meant by the term 'colinearity'?

17.20 In which order are the genes in the *HOM* cluster switched on? If one of the genes toward the 5′ end of the cluster is 'knocked out', the structures which it controls are said to be 'anteriorised'. Explain, with examples, what is meant by this term, and why does it occur?

17.21 Homeobox genes are strongly conserved throughout the animal kingdom. How was this found to be the case? Is this conservation consistent with Haeckel's 'biogenetic law'?

17.22 How many homeobox genes are found in mammals? On how many chromosomes are they located? What is their collective name?

17.23 Explain what is meant by a 'paralogous group'. How many paralogous groups are found in mammals?

17.24 Referring to Figure 17.8, indicate which genes in the mouse *Hox* cluster are paralogues of *C. elegans ceh-13* and *Drosophila lab*. Which mouse genes are paralogues of *Drosophila Antp*?

17.25 What is a rhombomere and how are rhombomeres related to other structures in the head region?

17.26 How are the *Hox* clusters related to the rhombomeres? How may this relatedness be investigated?

17.27 What evidence is there that the mammalian forebrain has a segmental origin? What are these putative segments called? Explain how investigations using the *Drosophila* embryo have provided useful insights.

17.28 What does the abbreviation '*unc*' stand for? In what organism have *unc* genes been investigated?

17.29 What do the letters in the POU acronym stand for?

17.30 With the aid of a diagram describe the POU-domain found in POU proteins.

17.31 Where are POU proteins located early in the development of the mammalian CNS.

17.32 Give some examples of how POU proteins are involved in controlling the differentiation of neurons within the CNS.

17.33 Show how the transcription of the *Drosophila Ddc* gene is controlled by interactions between two POU proteins. Compare this interaction with the Fos and Jun control of structural genes within a pinealocyte (review Section 16.7).

17.34 Describe the two superclasses into which homeobox genes may be classified.

17.35 It is found that the majority of early transcription factors act in a negative sense, i.e. they switch genes off rather than switch genes on. Why do you think they act in this way?

18

EPIGENETICS OF THE BRAIN

In this chapter we follow on from the genetics of Chapter 17 to consider how the cells of the brain differentiate and find their final places in the multi-billion unit system.

18.1 Distinguish between preformationism and epigenetics. Explain why preformationism is not an option.

18.2 What is meant by a morphopoietic or morphogenetic field?

18.3 Where is the proliferative zone located in the early vertebrate embryo? Describe the cellular dynamics of the cells destined to be neurons.

18.4 With which glial cells are neurons associated in their migration from the proliferative zone to the pial surface in the formation of the cerebral cortex? Describe some experiments that show the close symbiosis between neurons and glia at this early development stage.

18.5 What is the mouse weaver mutation (wv) and how does it help us in our attempt to understand the morphogenesis of the mammalian brain?

18.6 Discuss the importance of 'differential stickiness' of neural membranes in neuroembryology. Is there evidence that cells which originate together in the proliferative zone (or elsewhere) 'stick' together in later life?

18.7 What neuroanatomical defect is responsible for the defective behaviour of the mice carrying the reeler (rl) mutation?

18.8 On which chromosome is the reeler gene located? What is the name of the protein the gene encodes and what is believed to be its function?

18.9 Describe techniques by which the lineage of cells from the proliferative zone to their final location can be followed.

18.10 Approximately how many ommatidia constitute one of *Drosophila*'s compound eyes?

18.11 Describe a single ommatidium. How many retinula cells does it possess? What is so special about the seventh retinula cell (R7)?

18.12 What is lacking in the *Drosophila* sevenless (*sev*) mutant and how may this lack be detected?

18.13 What does the acronym 'Boss' stand for? What is the function of the Boss protein?

18.14 Describe the biochemical system which leads to the differentiation of R7. Does this system hold any lessons for the differentiation of cells in the mammalian brain?

18.15 What are 'microspikes' and what is their role in neuroembryology?

18.16 What is the difference between a 'neurotropic' and a 'neurotrophic' factor? Do some molecules have both functions?

18.17 Describe experiments which show the neurotropic action of nerve growth factor (NGF).

18.18 Which of the following statements is correct:

(a) Developing sympathetic fibres are selectively destroyed by anti-NGF antibodies.
(b) Developing parasympathetic fibres do not respond to the tropic influence of NGF.

18.19 List some important contact guidance molecules.

18.20 To which large group of molecules do the majority of neural cell adhesion molecules (N-CAMs) belong?

18.21 Describe the difference between N-CAMs and Ng-CAMs.

18.22 Draw a diagram showing the structures of N-CAM, Ng-CAM, MAG and Po. Which cells present these adhesion molecules in their membranes?

18.23 Describe the neuropathology which results from the mutation of the 11-N-CAM gene.

18.24 Draw a diagram to illustrate how N-CAMs from adjacent cell membranes interact with each other.

18.25 What is the molecular basis of the E to A conversion? Why does it not also occur in Ng-CAMs? What significance does it have in the development of the brain?

18.26 In most regions of the CNS more cells are generated than ultimately survive. Why? Give examples.

18.27 Describe experiments which show that cell loss at the centre is affected by target structures at the periphery.

18.28 Explain the difference between tropic and trophic factors.

18.29 Give examples of the trophic action of NGF on adrenergic and cholinergic cells.

18.30 How has a knowledge of the structure of NGF allowed investigators to identify other neurotrophic factors?

18.31 Describe experiments that suggest that some neurotransmitters can act as trophic factors. Give examples of the involvement of glial cells.

18.32 What does the acronym TGF-β stand for? Describe experiments that suggest that this molecule has a role as a growth factor in the developing brain.

18.33 To which part of the brain do optic nerve fibres in the lower vertebrates project? (See also Section 1.2.)

18.34 What is Sperry's chemoaffinity hypothesis? Describe some of the experiments that support this hypothesis.

18.35 Sperry's hypothesis has been subjected to fundamental criticism. Describe some of the experiments that suggest that it cannot be whole story.

18.36 With reference to Sections 18.8 and 18.10 say what is meant by the term 'critical' or 'sensitive' period.

18.37 With special reference to the mammalian primary visual cortex, define a 'receptive field' (RF) and an 'aggregate field'.

18.38 Explain how neurophysiologists discovered the existence of orientation and ocular dominance columns in the cat's primary visual cortex.

18.39 Describe some histochemical techniques which confirm the neuroanatomical reality of cortical columns. (See also Section 1.5 and Figure 1.23.)

18.40 Review experiments using the cat visual system which support the view that at least during the sensitive period the CNS is labile and open to environmental moulding.

18.41 How have experiments with the kitten ocular dominance system and the 'three-eyed' frog supported the hypothesis that synapses which fire together stabilise each other?

18.42 We noted in Section 14.4 that the neuromuscular junction (NMJ) has been called the *E.coli* of the nervous system. How has it proved valuable in analysing the hypothesis that synapses which fire together stabilise each other? Explain why application of the neurotoxin α-BuTX to all endings in an *en grappe* complex causes no atrophy.

18.43 In the context of the NMJ, what is meant by the term 'retrograde messenger'? Why do the experiments described in Section 18.9 imply the existence of such a messenger?

18.44 Describe experiments which suggest that the NMDA receptor is involved in the mechanism by which synchronously firing synapses are stabilised in amphibian optic tectum and mammalian primary visual cortex.

18.45 Discuss the balance between 'nature' and 'nurture' in the development of vertebrate brains.

19

MEMORY

In this chapter we examine aspects of the molecular biology of memory. The strategy, after some initial definitions, is to move from an analysis of the phenomenon in the simpler nervous systems of invertebrates to an application of the concepts derived from this analysis to the highly intricate brains of mammals.

19.1 How would you define 'memory' and 'learning'?

19.2 Define 'habituation' and 'sensitisation'. Provide examples of both phenomena.

19.3 In Pavlovian conditioning does the neutral stimulus precede or follow the unconditioned stimulus?

19.4 What is meant by the 'ISI'? What length of ISI has been found optimal?

19.5 Explain (with examples) what is meant by 'reinforcement' and 'extinction'.

19.6 Explain the difference between Pavlovian (or classical) and Skinnerian (or operant) conditioning.

19.7 In what type of conditioning is the reward 'contingent' upon the behaviours?

19.8 Give some examples of short-term memory (STM) and long-term memory (LTM). Why is STM sometimes referred to as 'dynamic' and LTM as 'structural'?

19.9 What is retrograde amnesia (RA)?

19.10 Describe some experiments whose results are consistent with Lashley's concept of 'mass action' or 'equipotentiality'.

19.11 Distinguish between 'implicit' (declarative) and 'explicit' (procedural) memory.

19.12 Review the neuroanatomy of the eye-blink reflex, especially Figure 19.4. On which cells in the cerebellum do signals from the conditioned and unconditioned stimuli converge? Referring back to Section 18.9 (stabilisation of synchronously firing synapses), suggest a molecular basis for establishing the conditioned reflex.

19.13 Review Box 9.2 and suggest how activation of mGluRs might be involved in the stabilising synapses which fire together on Purkinje dendrites.

19.14 Does loss of LTD after mGluR1 'knock-out' from Purkinje cells support any hypotheses you have developed in answer to Questions 19.12 and 19.13?

19.15 Enumerate reasons for believing that the mammalian hippocampus is essential to short-term memory.

19.16 Which invertebrates have proved valuable in the study of memory? Why and how?

19.17 Describe (with the help of a diagram) an experimental apparatus that has proved valuable in studying conditioned reflex and learning in *Drosophila*.

19.18 Numerous behavioural mutants of *Drosophila* have been isolated. In which of the three mutants (*dnc*, *Ddc* and *rutabaga*) mentioned in the text is:
(a) the mutation located on the X chromosome?
(b) the Ca^{2+}-dependent adenylyl cyclase affected?
(c) the cAMP-PDE affected?
(d) the synthesis of dopamine and serotonin affected?
(e) learning affected but not recall?

19.19 Review CREB proteins in Sections 16.6 and 16.7 and Question 16.9. Describe an experiment using *Drosophila* which is consistent with the hypothesis that CREB proteins are crucial for consolidation from STM to LTM.

19.20 Describe experiments which elicit classical (Pavlovian) conditioning in the sea hare, *Aplysia californica*.

19.21 In an experiment involving habituation of *Aplysia* gill withdrawal, it was shown that the EPSP in the motor neuron to the gill diminishes. Is this due to an alteration at the presynaptic or postsynaptic membrane? What is this phenomenon called?

19.22 Is it possible to study the monosynaptic reflex arc from *Aplysia* siphon to gill *in vitro*?

19.23 What is heterosynaptic facilitation? What neurotransmitter is involved at the presynaptic terminal in *Aplysia* heterosynaptic facilitation?

19.24 Review Section 10.2.3 and Table 10.1. Show how (in the *Aplysia* preparation) serotonin initiates a biochemical cascade which leads to an increased exocytosis of neurotransmitter from the presynaptic terminal of the sensory neuron.

19.25 Which of the following statements is correct:
(a) short-term sensitisation is blocked by transcriptional & translational inhibitors
(b) long-term sensitisation is blocked by transcriptional & translational inhibitors
What is the theoretical significance of these experiments?

19.26 Explain the similarity at the molecular level between long-term sensitisation and associative learning.

19.27 Define 'post-tetanic potentiation' (PTP) and 'long-term potentiation' (LTP). Is it the presynaptic or the postsynaptic membrane which is affected?

19.28 Explain how NMDA-R antagonists have been used to analyse the molecular basis of LTP in the mammalian hippocampus.

19.29 What evidence is there that the CREB site is involved in LTP?

19.30 Give an account of the dendritic spine. How may it be involved in the plastic changes underlying memory and what evidence is there that this actually occurs?

19.31 Review some of the many theories that have been proposed as the molecular basis of memory. Can you detect any underlying unifying principles? Do you think that the same molecular mechanisms are at work in invertebrates and vertebrates and in different types of memory in vertebrates?

20

SOME PATHOLOGIES

One of the major reasons for studying neurobiology is to help ameliorate the human condition. Neuropathologies, especially the neurodegenerations of old age, present some of the last great challenges to medical science. In this chapter we see how molecular approaches are beginning to yield fundamental understandings and how tiny molecular changes multiply up through the systems of the brain to yield devastating results.

20.1 Distinguish between neurological and psychological illness. Do you think the distinction is sharp? Useful?

20.2 Review Box 8.1 and Sections 9.4 and 15.5 and explain the dangers of markedly increased levels of internal Ca^{2+} on cerebral neurons.

20.3 List the major human and non-human prion diseases.

20.4 What test would you use to establish the chemical difference between a virus and a prion?

20.5 Write down abbreviations used in referring to prion genes and proteins.

20.6 On which chromosome is the CJD gene carried? What is believed to be the biochemical alteration caused by mutation of this gene? How does this alteration transform the prion protein into a pathological agent?

20.7 How does replication of a prion protein differ from that of a nucleic acid?

20.8 How has the transmission of prions through the brain been studied? Which parts of the brain are most affected?

20.9 Discuss the evidence that prions are transmissible between different species.

20.10 Review Box 3.1 and explain how antisense oligonucleotide therapy might prove to be an effective therapy against prion diseases.

20.11 What is the metabolic defect in PKU? As a consequence of this defect, what metabolite is present in excess?

20.12 Why is PKU seldom found in people of African descent?

20.13 What therapy can be employed to control the symptoms of PKU?

20.14 On which chromosome is the FMR1 gene located? What peculiarity in the base sequence is believed to be responsible for the FraX syndrome? Why is this peculiarity suspected of being responsible for the increased severity of the disease often observed in successive generations?

20.15 Distinguish between the symptoms of NF1 and the NF2 neurofibromatoses.

20.16 How does the NFI gene product (neurofibromin) resemble a GTP-activating protein (GAP)?

20.17 Referring to Box 7.1 and Figure 20.3, explain how mutations affecting the GRD of neurofibromin can cause cancerous growth.

20.18 Why are the choreas sometimes called 'inverse Parkinsonisms'?

20.19 What are RFLPs and how are they prepared?

20.20 On what chromosome is the gene for HD located? How do we know? What is its name? How does it resemble the FMR1 gene?

20.21 How has the Old Order Amish community assisted in the study of endogenous depression?

20.22 Review Section 15.4 and outline the roles of neurotransmitters in causing depression. Indicate how some antidepressants work.

20.23 Discuss the difference in time-scales between the pharmacological action of antidepressants and their psychological consequences. (Review Section 9.1.3 and the paragraphs on the NMJ in Section 18.9 for examples of the dynamics of receptor turnover.)

20.24 Describe the symptoms of Parkinson's Disease (PD). What is meant by 'festination'?

20.25 Which brain areas are principally involved in PD? Where is the substantia nigra located in the brain?

20.26 With the aid of a diagram, describe the neural loops controlling movement which may be affected during Parkinsonism. Explain how a delicate balance of inhibition and disinhibition in these loops normally controls output to the primary motor cortex.

20.27 Which neurochemical system is principally affected in PD? How does this affect the operation of the loops described in Question 20.26? How may this defect be treated?

20.28 Which serendipitous discovery increased our insight into the molecular causes of PD?

20.29 What is MPP^+, how and where is it formed, and how does it affect neurons?

20.30 Is there any evidence that PD may have an environmental cause?

20.31 Do you think that transplant surgery holds hope for a future therapy? Give reasons and examples in your answer.

20.32 Give an account of the symptoms of Alzheimer's disease (AD).

20.33 Describe the three 'hallmark' lesions of AD neuropathology

20.34 What is amyloid and where is it found?

20.35 Describe the relationship between APP, β-A4 and amyloid.

20.36 On which chromosome is the APP gene located and why is this significant?

20.37 Discuss the isoforms of APP. What is the significance of the sequence of 56 amino acid residues following residue 289 in APP_{751}?

20.38 Is the APP gene the only candidate gene in AD? If not, which other genes are suspected and on which chromosomes are they located?

20.39 What is the biochemical function of ApoE and how do its three isoforms differ? Which isoform has been associated with AD and what functional rationale can be proposed for this association?

20.40 Are any environmental agents suspected of being implicated in AD? If so, say what they are and assess the evidence for this putative association.

20.41 With the aid of a diagram, outline the amyloid cascade hypothesis. What future discoveries would disconfirm this hypothesis?

20.42 With reference to the increasing genetic predictability and present incurability of the neurodegenerations, discuss the proposition ‘it is folly to be wise’.

APPENDIX 1

MOLECULES AND CONSCIOUSNESS

In this appendix we lift our eyes from the detail of molecular neurobiology to the broad (and deep) perspectives of one of the crucial issues in neuroscience and philosophy: consciousness in a world of things.

A1.1 Discuss what is meant by the term 'consciousness'?

A1.2 Provide some examples of 'raw feels' or 'qualia'.

A1.3 Distinguish between the 'easy' problems and the 'hard' problem of consciousness.

A1.4 Why is it important to have some answer to the question of how far through the evolutionary tree consciousness is found?

A1.5 What is the 'Chinese room' argument?

A1.6 Discuss the notion of levels of complexity in the brain. How many such levels would you recognise, and why?

A1.7 With reference to Box 13.1, discuss Emil du Bois Reymond's distinction between questions to which we can answer *ignoramus* and those to which we answer *ignoramibus*. Compare du Bois Reymond's classification with that of Question A1.3. Which analysis do you prefer?

AFTERWORD

CURRENT, VOLTAGE, AND ALL THAT

The brain is, among other things, a processor of electrical signals. Nerve fibres carry impulses that are electrically detectable as action potentials. Neural membranes contain highly adapted macromolecules that act as sophisticated electrical devices. Therefore the study of electricity is fundamental to neurobiology. This Afterword is intended to help you review the basics.

Because all matter, whether steel, wood, seawater or brain tissue, consists of particles moving in a vacuum, an analysis of the interactions of particles in a vacuum should be sufficient to describe all material phenomena. And so it is, in theory – although this description requires some mathematical sophistication to be fully appreciated. However, matter is too complex and dense to be conveniently describable in terms of simple two-particle interactions. As a result, another approach was developed, in which a few particles are singled out for attention, the remaining particles being relegated to the role of the *medium*. This allows us to write simple, workable equations, but with one hitch: the properties of the medium have to be described by special empirical laws. When these laws are simple proportionalities, the equations are *linear* and reasonably easy to solve; otherwise, the equations are *nonlinear*, requiring special methods for their solution.

Since linear laws are straightforward to write down and apply, it makes sense to try them first. Thus, in biology as well as in other fields, linear laws were applied initially. Unfortunately, the predictions of these laws sometimes do not fit the data, which can lead to complications (just as in early astronomy, the attempt to describe planetary motions as circles led to an increasing series of epicycles). It may be that some biological matter is inherently nonlinear. These nonlinearities may even be responsible for such important phenomena as the opening and closing of ion channels (see Box 10.1 on page 234).

CHARGES, FORCES AND FIELDS

Matter consists of atoms, and each atom consists of a positive nucleus in a cloud of negative electrons. The terms positive and negative describe the electric *charge* carried by these particles. Electrical forces hold the atom together and are involved in the bonding that forms molecules. Charges at rest give rise to electrostatic forces. The force between two charges, each concentrated in a small region of space (point charges), is a vector directed along the line between them. It is repulsive if the charges are of like sign and attractive if they are of unlike signs. Quantitatively, the magnitude of the force is proportional to the size of each charge, and inversely proportional to the square of the distance between them, a relation known as Coulomb's law. The unit of charge is the coulomb (C), which is equal to 6.24×10^{18} proton charges. The charge of a proton is 1.60×10^{-19} coul, and the charge of a mole (6.02×10^{23}) of protons is the faraday (F), about 96 500 C.

Although Coulomb's law seems to suggest that one charged particle acts on another charged particle *directly*, it is valid to say that the first particle alters the space around itself in a certain way, and the second particle responds to that alteration. The first particle (the source) sets up a *field*, and the second particle (the test particle) is acted upon by the field. The field is real. When the source particle is moving (as, for example, an electron on an antenna) the field has a magnetic as well as an electric component, and so is called an electromagnetic field. Radio waves, light and X-rays are electromagnetic fields; these travel at the speed of light.

One convenient way to set up an electrostatic field is with two parallel metal plates with opposite charges on them. No matter where we place a small test particle between the plates (except near the edges), the force will be the same. Thus we can say that the field between parallel plates is uniform (assuming the medium within is uniform).

ELECTROSTATIC ENERGY AND VOLTAGE

To move a particle through a field takes work. To move a positive test particle closer to the positively charged plate, we have to do positive work, expending energy. If we release the particle, it will 'fall" toward the negatively charged plate, giving up energy to heat, kinetic energy or some other form. Since work has had to be done in separating the charged plates (or charging the plates), this device stores energy. It is called a *capacitor* or condenser. Charging the capacitor takes work; discharging it yields energy. The symbol for a capacitor used in circuit diagrams is shown in Figure 1a. The unit of work or energy is the joule (J); $1\,\text{J} = 1\,\text{kg}\,\text{m}^2/\text{s}^2$.

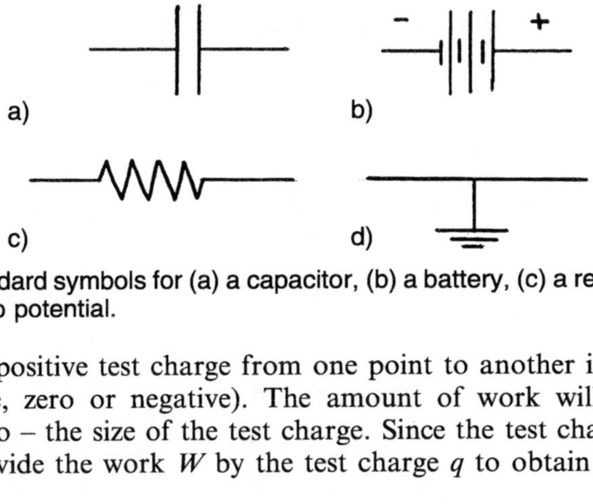

Figure 1 Standard symbols for (a) a capacitor, (b) a battery, (c) a resistor and (d) a point at earth (ground) or zero potential.

To move a positive test charge from one point to another in an electric field requires work (positive, zero or negative). The amount of work will depend on – in fact, is proportional to – the size of the test charge. Since the test charge itself is usually not of interest, we divide the work W by the test charge q to obtain the *potential difference* or *voltage* V:

$$V = \frac{W}{q}$$

The unit of voltage is the volt; its symbol is (also) V; $1\,\text{V} = 1\,\text{J/C}$. A convenient source of voltage is a battery, a device that converts chemical energy to electrical energy. The standard symbol for a battery in circuit diagrams is shown in Figure 1b. Voltage is also produced by (among other things) electrical generators, solar cells and the electric organs of certain fishes. A difference in ion concentrations can produce a potential difference known as an electromotive force (emf); see Chapter 11.

We now return to the capacitor to define its effectiveness in storing charge. The *capacitance* is the amount of charge stored per unit potential difference:

$$C = \frac{Q}{V}$$

The unit of capacitance is the farad (F); 1 F = 1 C/V. If the space between the plates is vacuum (or air), the capacitance of a capacitor with area A and plate separation d is given by:

$$C = \frac{\varepsilon_0 A}{d}$$

The units for d are metres; A is in square metres, and Q is in coulombs. The factor $_0$ is a universal constant, equal to $8.854 \times 10^{-12}\, C^2/N\,m^2$.

If we now place some material between the plates, the force on a charge will be decreased by a factor of ε, called the dielectric constant or (more correctly, since it is not always constant) the *dielectric permittivity*. The energy stored will be similarly decreased, while the capacitance will be increased by that factor:

$$C = \frac{\varepsilon\varepsilon_0 A}{d}$$

The dielectric permittivity is 1 for vacuum (or air), about 2 for oil and about 80 for water. The equation shows that the larger the area, the larger the capacitance; the larger the plate separation, the *smaller* the capacitance. In this way the myelin sheath, which separates the inner and outer conducting solutions of an axon by a far greater distance than the plasma membrane alone, has the effect of decreasing the capacitance of an axon.

CURRENT

A mobile charge will move, if positive, from higher to lower electric potential, and if negative, from lower to higher. An ongoing movement of charges is called a current. Quantitatively, *current* is defined as the time rate of movement of charge; the direction of the current is that of the positive charge. If an amount of charge Q traverses an area in a time t, the current I is given by:

$$I = \frac{\Delta Q}{\Delta t}$$

The unit of current is the ampere, abbreviated A; 1 A = 1 C/s. A material that supports the flow of current is called a conductor; one that blocks current is an insulator (or dielectric).

We know that metals (elements on the left side of the periodic table) are good conductors because their valence electrons are only loosely bound to the core of the atom (the nucleus and inner electrons). It takes little energy to put these electrons into a state (the 'conduction band') in which they are free to move through the metal. Electrons bounce in a random pattern through the metal; when a voltage is present, a *drift* up the voltage gradient is superimposed on that random pattern.

Another type of conductor is an electrolytic solution, such as salt water. To apply a voltage, two conducting electrodes are inserted into the solution, a cathode (negative) and an anode (positive). The net current is the sum of the currents carried by each type of ion. The applied voltage causes the positive ions (cations) to drift toward the cathode and the negative ions (anions) to drift toward the anode. The drift speed for a given applied voltage differs according to the type of ion and the solvent. The drift speed per unit voltage is called the *electric mobility*.

Ionic conduction also occurs in certain solids, called *superionic conductors*, when the conduction rate is high. In contrast to electrolytic solutions, which are relatively unselective, superionic conductors only conduct specific sets of ions.

OHM'S LAW

The application of a potential difference to a material will cause any mobile electrons in it to drift, producing a current. On the other hand, the passage of a current through a material will redistribute the electrons or ions in it, leading to a potential difference. Thus a voltage causes a current, and a current causes a voltage. In many materials, the current will be directly proportional to the voltage. Such materials are said to obey *Ohm's law*:

$$I = \frac{V}{R}$$

where R is a constant and is called the *resistance* of the material.

This equation can be interpreted in two ways. One is that it defines R:

$$R = \frac{V}{I}$$

This equation is always true, because it is the definition of R, but it is not a law. For it to be Ohm's law, we must specify that R is a constant. Whether the resistance of a material is constant or not has to be determined by experiment. For a metal, Ohm's law holds over a wide range of temperature, but not when the wire becomes incandescent. It does not hold for devices such as rectifiers and thyristors. Ohm's law applies to electrolytic solutions, but the electrodes may not be 'ohmic'.

The unit of resistance is the *ohm*; 1 ohm = 1 V/A. Frequently used in electrophysiology is the *conductance*, the reciprocal of the resistance, expressed as $g = 1/R$. The unit of conductance is the siemens; 1 S = 1 A/V. (The obsolete term, mho, appears in older works.) The symbol used in circuits for resistance or conductance is shown in Figure 1c.

The resistance of an object such as a piece of wire or membrane depends on its dimensions. It is useful to obtain a quantity that is independent of dimensions but depends only on the nature of the material (as well as its temperature and pressure). If we lay two similar pieces of uniform membrane side by side and connect their surfaces to a voltage source, the area is doubled and the current will be twice as much as for one. Since the voltage is unchanged, we conclude that the resistance is halved. More generally, the resistance is inversely proportional to the area A perpendicular to the direction of current. If instead we stack the two membranes one on top of the another, the area is unchanged but the thickness is doubled. To obtain the same current as before, the voltage must be doubled (twice as much work to push a charge across two membranes as one), so the resistance is doubled. Thus the resistance is directly proportional to thickness or length L. Putting the statements together, we see that:

$$R = \frac{\rho L}{A}$$

where ρ is a proportionality constant called the resistivity. It is expressed in ohm-metres. Its reciprocal is the conductivity, given in units of S/m.

When we are interested in the resistance of a membrane without specifying its size, the relevant quantity is resistance times area, RA. The units are ohm m^2. This quantity is useful, for example, in studying the cable properties of an axon (Chapter 11). Its reciprocal is the conductance per unit area, in S/m^2.

CIRCUITS AND MEMBRANE MODELS

Circuit diagrams use conventional symbols for capacitors, batteries, resistors, switches, amplifiers and other circuit elements, connected by ideally nonresistive wires; another useful symbol is that for *earth* (*ground*), the point at which the electric potential is considered to be zero (Figure 1d). Elements connected end-to-end are said to be *in series* (Figure 2a), while elements with their terminals connected together at both ends are said to be *in parallel* (Figure 2b). Specialised courses deal with the behaviour of circuits under direct-current, alternating-current and transient conditions.

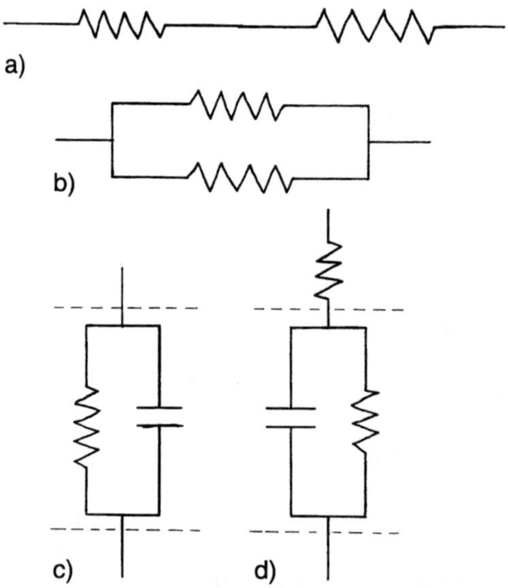

Figure 2 (a) Two resistors in series. (b) Two resistors in parallel. (c) A resistor and a capacitor in parallel. This represents a crude model of a biomembrane. (d) A parallel resistor–capacitor combination in series with another resistor; this may represent a membrane together with its access resistance, the resistance of the material between the electrode and the membrane. In the case of the squid axon, a layer of Schwann cells contributes to the access resistance.

Circuits are of interest to us here not only because they constitute the basis for electrical measurement techniques such as the voltage clamp, but also because membranes and their components, such as bilayers, channels and receptors, are frequently modelled by combinations of circuit elements. The bilayer of an excitable membrane may, with reasonable accuracy, be modelled as a capacitive element. The ion-selective channels have been modelled as variable conductances, each in series with the emf due to the concentration difference of its particular ion. These models are useful because they give us a concrete object to analyse, but we must remember that such a model represents a set of assumptions. The results of these analyses can only tell us the behaviour of the model, not necessarily that of the object it is intended to model. The validity of any model can only be determined by comparison with experiment and the application of Occam's razor.